6 STEPS TO EFFECTIVE WRITING IN SOCIOLOGY

D0166304

6 STEPS TO EFFECTIVE WRITING IN SOCIOLOGY

Judy H. Schmidt
Harrisburg Area Community College

Michael K. Hooper
California Department of Justice

Diane Kholos Wysocki
University of Nebraska – Kearney

WADSWORTH
CENGAGE Learning

Australia • Brazil • Japan • Korea • Mexico • Singapore • Spain • United Kingdom • United States

WADSWORTH
CENGAGE Learning™

6 Steps to Effective Writing in Sociology
Judy H. Schmidt,
Michael K. Hooper and
Diane Kholos Wysocki

Publisher: Sabra Horne

Development Editor:
  Shelly Murphy

Assistant Editor:
  Stephanie Monzon

Technology Project Manager:
  Dee Dee Zobian

Marketing Manager:
  Matthew Wright

Marketing Assistant: Michael
  Silverstein

Project Manager, Editorial
  Production: Candace Chen

Print/Media Buyer:
  Rebecca Cross

Cover Designer: Natalie Hill

© 2003 Wadsworth, Cengage Learning

ALL RIGHTS RESERVED. No part of this work covered by the copyright herein may be reproduced, transmitted, stored, or used in any form or by any means graphic, electronic, or mechanical, including but not limited to photocopying, recording, scanning, digitizing, taping, Web distribution, information networks, or information storage and retrieval systems, except as permitted under Section 107 or 108 of the 1976 United States Copyright Act, without the prior written permission of the publisher.

For product information and
technology assistance, contact us at **Cengage Learning
Customer & Sales Support, 1-800-354-9706**

For permission to use material from this text or product,
submit all requests online at **www.cengage.com/permissions**
Further permissions questions can be e-mailed to
permissionrequest@cengage.com

ISBN-13: 978-0-534-17290-9

ISBN-10: 0-534-17290-3

Wadsworth Cengage Learning
20 Davis Drive
Belmont, CA 94002-3098
USA

Cengage Learning is a leading provider of customized learning solutions with office locations around the globe, including Singapore, the United Kingdom, Australia, Mexico, Brazil, and Japan. Locate your local office at **www.cengage.com/global**

Cengage Learning products are represented in Canada by Nelson Education, Ltd.

To learn more about Wadsworth, visit
www.cengage.com/wadsworth

Purchase any of our products at your local college store or at our preferred online store **www.cengagebrain.com**

Printed in the United States of America
7 8 9 10 11   19 18 17 16 15

TABLE OF CONTENTS

PREFACE

Conceptually, effective writing is the ability to convey clearly thoughts on paper (or on a computer screen) for the benefit of an intended readership. *6 Steps to Effective Writing in Sociology* develops a standard method for approaching writing tasks, whether academic or professional, that is both logical and efficient to produce writing that reflects purpose, audience, and scope. Writers are guided through a series of specific thinking and writing skills that writers can apply at each step of the writing process: thinking about writing, building structure, preparing to write, drafting, and revising.

The authors of this book have extensive teaching experience and have written for academic and professional audiences. Professor Schmidt is an English instructor experienced in providing writing instruction to students and consultation to faculty and practitioners. Dr. Hooper is a practitioner who has prepared and reviewed the gamut of professional (and academic) reports and has taught both college students and criminal justice practitioners. Dr. Wysocki is a Sociology professor who is the author of a research methods reader and has extensive experience in writing and publishing sociological research papers.

The authors have reviewed thousands of papers and reports and thus bring with them extraordinary in-depth knowledge of all facets of academic and professional writing. Their expertise enables them to recognize the common impediments to effective writing and at the same time to provide the corrective guidance that leads to clear written communications. In essence, the authors have isolated the fundamentals of good writing and presented them in a uniquely illustrative manner.

6 Steps to Effective Writing in Sociology is *not* intended to be a grammar or usage text, as it barely addresses these topics which are readily available in myriad technical texts. Instead, this book's value is its service as a guide for *approaching* any writing assignment. The compact handbook's logical and conversational flow enables the user to page quickly through its succinct content and extract the pointers that are pertinent to the task at hand and quickly fashion a responsive document.

The authors wish to acknowledge the following reviewers: Michael Lovaglia, University of Iowa; Mark Packard, Southeast Community College; and Steven Vassar, Minnesota State University, Mankato.

STEP ONE :
THINKING ABOUT WRITING PROJECTS

GETTING STARTED

Imagine this. You are sitting in a classroom, hearing your instructor begin, "I would like you to write a paper...." She then goes on to mention specifics about the assignment requiring research and strong writing skills. Perhaps your first reaction is to feel somewhat overwhelmed and apprehensive, trying to decide how you will handle the assignment--and maybe certain that you can't do it!

In case you imagine you are the only person who reacts this way when asked to write, rest-assured you have lots of company. In fact, many Sociology majors and professionals in the field indicate they are uncomfortable knowing how to approach required academic and professional writing tasks. Because they are unaware of an efficient, effective manner for thinking and then writing, each assignment seems like a new test where no guidelines exist.

Taken to the extreme, some beginning writers may believe that good writing is genetic, and, if you're not born with it, you can't be a strong writer. They don't understand that good writing consists of learning specific skills and then applying them to the writing situation after they decipher the writing problem, not unlike learning to play golf, using a map to drive to a new location, or determining the murderer in a crime novel.

Therefore, the purpose of this compact resource manual is to help develop strong writing skills for you as Sociology students that can be used now in academic classes and later in professional writing. It develops a series of specific writing and thinking skills to be applied at each step of the writing process: thinking about writing, building structure, preparing to write, drafting, and revising. While many examples use Sociology-specific topics, all writing skills can be readily applied to other courses and professional writing situations. Each paper assignment you have will require you to do a number of the tasks described in this book. It is important to address all of the specific tasks required by your teacher and the assignment given.

FOLLOWING THE PLANNING CLUES: TASK, SCOPE, AND READERS

TASK

In our scenario above, when the beginning writer was feeling tense, a more confident, experienced writer was also listening with interest to the assignment. The major difference was that this student identified clues in the instructor's assignment, providing helpful information for the thinking about, planning for, and writing of the paper. The clues are: **task, scope, and readers**.

The **task** word(s) tells writers what writing action they will take with the information they have found which helps writers develop and reject ideas based on the assigned task.

Looking at the list of common task words following, note how each requires the reader to think about the information differently.

2

Task	Thinking Pattern
Analyze	break into parts
Compare	identify similarities
Contrast	identify differences
Critique	express opinion about specific aspects
Define	give precise meaning; identify specific parts
Develop	expand upon, elaborate
Evaluate	appraise, judge
Explain	give reasons for, make clear
Identify	choose, select
Illustrate	give examples of
Interpret	explain the meaning or significance
Review	reexamine, look over
Summarize	condense; go over main points only

Notice that each task word directs the writer to different writing purposes and cannot be used interchangeably. If your instructor has written comments such as "unfocused," "didn't answer the question," or "confused purpose," chances are great that you did not meet the assignment's **task.**

SCOPE

After the writer has a basic understanding of the kind of thinking the paper will require, the **task**, it is important to consider another clue, the **scope**, which is the amount of

3

material that would need to be covered to complete the writing tasks.

Consider these factors when determining **scope**:

- number and kinds of writing tasks in assignment
- required number of pages or words
- difficulties in finding research
- reader's knowledge of subject area

The first two factors, writing task and paper length, are easy to understand in terms of choosing a subject topic. But writers often miscalculate the scope, resulting in a subject topic that is too limited, leaving the writer with a lot of extra space to try to fill, or too broad, forcing the writer to cover material very superficially because there isn't time to cover it adequately. **Remember, not every subject topic fits every assignment.**

The third factor, difficulties in finding research, asks the writer to anticipate difficulties in finding the necessary research to develop the topic. Is the topic so broad that the writer would need months to gather necessary information? Is there very little research that could be done because the topic is too vague and undefined? Would the writer need some special permission to acquire research from sources? (Think government, military, personnel files, ongoing research here.) Is the topic so new that journal articles, or at least the ones most college/public libraries subscribe to, are not yet available?

Readers' knowledge of the subject is also a determining factor in choosing a subject topic. If identified readers know very little about the subject, the writer may have to "spend" a large number of pages to bring them to the "starting point" of the paper. Conversely, beginning writers

4

may want to avoid a subject area when the readers are experts in the field, recognizing the writer's grasp of information is still too elementary to capture reader interest. Writers should also assess subject topics for emotional reactions they may produce in readers. Of special concern are those subject topics that produce very opposite but equally strong feelings/beliefs in readers or positions that the writer already disagrees with.

READERS

The last writing clue to be considered at this "thinking about" stage is to identify who will read your paper. Generally, in academic classes, the reader is the instructor unless otherwise specified, while professional sociological correspondence may have levels of readers with reports often being circulated upward or used as court documents.

Whether academic or professional assignments, writers need to identify their **readers** to answer three important questions:

- What does the reader already know about this topic?
- What information does the reader need to know in order to understand this topic?
- What preconceptions, if any, does the reader have toward my topic?

ANALYZING A SAMPLE ASSIGNMENT

Only when writers look at a paper assignment in terms of the task to be accomplished, the scope of the assignment, and the readers can they begin to generate appropriate ideas.

5

Let's see how this would work. Suppose your Sociology instructor gave your class this paper assignment.

SAMPLE ASSIGNMENT

The field of Sociology provides a wide variety of research opportunities, interesting to both students and practitioners. Therefore, the purpose of this paper is first to ask you to identify a research topic that provides recent information to sociological practitioners that would be helpful to them in their mission to find out about some aspect of social life. To help you select a current topic, please review current sociology academic publications and/or go through your textbooks to find an idea that interests you.

Then, after you have selected a writing topic, prepare a research paper that summarizes at least three key aspects of the current sociological area relating to the aspect of social life you have chosen that you judge practitioners would need to know in order to understand the information. The paper should be approximately ten pages, using at least five different research resources. All resources should be documented using ASA documentation format. The due date is October 21, and the paper is worth 30 percent of the course grade.

As an added incentive to perform well on this exercise, a review panel of sociology professors will read each paper to determine which three papers best address topics of urgent value to social life.

Keeping this in mind, let's go back to the sample writing assignment to analyze it in terms of writing clues:

Task(s)   identify and summarize

Scope   approximately 10 pages

Reader   your instructor and other sociology professors

GENERATING IDEAS FOR WRITING TOPICS

Here's the good news: you have come up with a subject topic, so now you're ready to write the paper. Here's the bad news: no, you're not!

To believe that, after choosing a subject topic, the writer is now ready to write the paper is a mistake that many beginning writers make. This kind of thinking is like deciding to drive to a new location several states away and just getting in the car and driving or deciding that you want to build a log cabin home and starting to build without blueprints. You may eventually have limited success, but certainly the frustration level will be high as you use trial and error to accomplish your goal.

A better plan is to understand that thinking through your subject topic enables you to make important choices about a subject that interests you, to break it into smaller components, to determine research and resources you want to use and are available, and to develop an organizational plan to determine the order that information will be included in the text. Only then are you ready to begin drafting.

It's very difficult to write a strong paper about a weak idea. How often have you heard a classmate say, "I didn't know what to write about. If the instructor had given me a topic, I could have turned in a good paper"? Yet, at this level of education, one of the tests of your understanding of course material is to be able to apply it, in part by determining a specific writing focus that allows you to show your mastery of course information.

GENERATING IDEAS FOR SAMPLE ASSIGNMENT

Looking again at our sample writing assignment, let's list several possible subject topics we could write about in sociological areas that relate to some aspect of social life:

1   terrorism
2   gangs
3   nutrition/exercise
4   racial profiling
5   technology

Because each listed subject is so broad, we need to come up with a method that allows us to break down the subject topics to more narrowed areas to see the smaller subject areas of each subject topic.

Several methods to explore the topic and narrow ideas for writing exist, including **brain drain**, **free writing**, **listing/outlining**, and **reading**. However, in order to any use of these techniques, the writer needs to consider **all** generated ideas before discarding any of them. Sometimes the best ideas are those that, at first glance, seem strange or unconnected to the assignment.

TECHNIQUES TO NARROW SUBJECT TOPICS

BRAIN DRAIN is a technique to discover what your brain may be thinking. First, write the subject topic on a sheet of paper; then take approximately 3-5 minutes to list everything you associate with the topic. Do not consciously think before writing or evaluate anything on the list until time is up. Then look at the list, put similar ideas together, but also consider those that have no connection to other items on the list. You may have identified a more narrowed topic.

FREE WRITING is similar to brain drain. Keeping the writing topic in mind, a writer writes for 5 minutes without stopping, recording everything that comes into mind, even ideas that are unrelated to the writing topic. If the writer is unable to think of anything, "I can't think of anything; nothing comes to mind," should be written until something pops into mind. The trick is to record all thoughts within the 5-minute period. When time is up, the writer reads what has been written, considers it, and circles any ideas that are more focused and of interest as a writing idea.

LISTING is a casual approach to discovering content with the writer recording ideas on a list to make frequent additions and deletions to it. The writer makes no attempt to evaluate the usefulness of what is listed or consider its relation to other listed items. This technique, often preferred by visual learners, allows the writer to keep track

9

of what may be very separate ideas to consider or to discover that many ideas actually fit together to form a central idea. One of the advantages of an idea list is that it is easily tucked into a pocket, allowing the writer to look at it frequently, whether in line in the cafeteria, waiting for the movie to begin, or sitting at a desk.

OUTLINING requires the writer to have a more structured idea of the writing subject and the major pieces of information that would be necessary to develop it. Therefore, some outlines are produced after writers have used the other techniques listed here. The writer can use a formal outline format with Roman numerals, letters and numbers or a topic outline form where the writer uses phrases to describe each section.

Who would opt for outlining? Certainly no one who feels the urge to begin writing right away because outlining requires the writer first to have a clear idea of writing direction. Writers in technical fields seem to choose this method that can create a "blueprint" writing plan, that is, everything decided and structured before drafting begins.

Don't let this scare you away from using it. The outline format provides a visual record that helps writers spot areas that are missing or weak, directing the writer to think through content and spot flaws in organization.

READING about the topic helps writers break down the subject topic into smaller parts because the reader often finds articles that deal with narrowed topics. And, if the narrowed topics are still large or complex, the writer may determine the subject topic may not be appropriate for the assignment task, scope, or readers. Reading has an additional function: it may show the writer how little he/she knows about the topic or challenge existing opinions/beliefs.

When a paper assignment is given, resist the impulse to begin writing the paper immediately. Instead, spend time analyzing and understanding the assignment, considering **task**, **scope**, and **readers** to help the writer identify possible writing topics.

Using this process is a good start in planning, but writing topics then need to be narrowed to develop a writing idea. **Brain drain**, **free writing**, **listing and outlining**, and **reading** are methods to help writers break the writing topic into smaller pieces of information to discover the writing idea for the assigned paper. In Chapter Two our listed subject topics are narrowed, using techniques discussed in this chapter.

It's natural for us to want to plunge into the writing project, immediately producing written pages, rather than spending time thinking and planning. However, by spending time understanding the writing project and giving ourselves time to generate and test ideas, we will save time later in the writing process, without trying to unravel writing snarls.

STEP TWO:
BUILDING THE PAPER'S
STRUCTURE

It would be so easy if only your instructor chose to tell you what to write about, which could eliminate so much thinking, much the same way you knew in grade school that you would begin each year's English class with the perennially exciting composition, "How I Spent my Summer Vacation."

Yet, since the purpose of college is to grow intellectually and to learn how to think critically to discover connections among information, writers first make a series of decisions about a paper's direction that builds the structure of the paper. Later it will be filled in with pertinent information and organized to meet readers' needs.

Three of the structural decisions this chapter looks at are: **narrowing the writing topic, gathering information from library sources, interviews and surveys, and then determining a tentative thesis statement.** (The tentative thesis statement will be mentioned in this chapter, but refer to "Step Three" for a more complete discussion.)

NARROWING THE SUBJECT TOPIC

In the first chapter we discussed techniques that allow the writer to both generate and then narrow subject topics into smaller chunks of information, all focused on social life. The problem is that the subject topics we generated in the first chapter are too broad and could not be handled in the scope of this writing assignment. Because we would have to skim over points too briefly to develop information, sort of a "drive-by" approach, fast and too busy to stop, we still need to narrow the subject topic to something we can develop intelligently, correlating to **task, scope, readers.**

Looking at the narrowed subjects below, you can see that we are beginning to move in more specific and focused directions. This in turn will give us a better idea of the research we will want to find.

Subject Topic	Narrowed Subject
Terrorism	children's perceptions of terrorism, charity efforts, tracking terrorism, social construction of terrorism, cost of terrorism
Gangs	violence, drug use and marketing, clothing, recruitment, urban gangs, rural gangs
Nutrition/Exercise	Stress reduction, exercise programs, college student dietary practices, gender differences, eating disorders

Racial Profiling	prevalence, media coverage, geographical patterns, negative impact, historical perspective, social aspects
Technology	distance education, attitudes towards technology, social aspects, computer crime against children, health, media coverage

Note: Even though we have narrowed the topic from very general to more specific, we still need to evaluate the narrowed topic in terms of the assignment. Because we are not being asked only to gather information **about** a topic, we must consider, "What about this narrowed topic connects to the idea of social life and sociology?"

GATHERING INFORMATION

Because we are not experts in the field, we don't know the answers and instead need to gather relevant information in order to decide if we're going in the right direction. The way to begin the process is by researching relevant information from library sources, and possibly interviews and surveys if you have taken methods courses and have the permission of your professor.

The **researched information** we find has three purposes:
1. it helps the writer continue to narrow and refine the subject topic
2. it helps the reader to determine how information fits together (the thesis statement)
3. it provides specific information to support and illustrate the paper's thesis statement

RESEARCH

It's easy to feel intimidated in the library, wondering where to look and what to look for. At this point, because we are exploring more general subject topics, we want to look at more general references from books, databases, periodicals, and on-line sources. Other than your professor, who is the best person to talk to about what to look for? The **reference librarian**, who probably has more suggestions about where to start than you have questions. Did you take the library tour during orientation? If not, or if you forgot, the reference librarian can show you where information is located and even how to use the computerized library systems.

The very first thing you need to know is just some basic terms regarding your sources. I find that it is best that students use mainly **primary sources**. A primary source is the original article that describes and reports **empirical research** and **theoretical articles** that have been published in a **peer-reviewed** academic journal. An empirical research paper is a description of a systematic observation or study, while a theoretical article explains a theory and shows how it relates to the variables being investigated. It's important that you use a peer-reviewed article because that means the research paper has been blindly reviewed, usually by three other scholars in the same field, who decide if the research is up to the standards of the journal.

Secondary sources, on the other hand, are just summaries of original research that you find in newspapers, magazines, and hear about on the news. Secondary sources give clues about where to go look for the original study, but it is difficult to know for sure that the reporter interpreted, and then reported, the results of the original research correctly. Think here of the difference between using the "Declaration of Independence" (primary) vs. a history book that discusses the importance of the document (secondary source). You might ask your professor if he/she wants you to use primary or secondary sources.

While you may be using more general reference resources, you should also be aware of special sociology resources. (Yes, the reference librarian knows about them, too.)

Some of the possible sources for reading in sociology academic journals, e.g., *Sociology* and *American Sociologist,* documenting the results of research on current issues of research to both practitioners and academicians. In addition, there are more specialized periodicals, e.g. *Gender and Society, Family Planning Perspectives*, and *Sex Roles*. There are numerous books on contemporary issues in Sociology as well

In recent years the federal government has subsidized numerous entities, e.g., *Centers for Disease Control, Statistical Abstracts of the United States,* and *Bureau of Labor Statistics*, which have as their principal mission research to further sociological problem solving.

 These research enterprises publish their findings in a wide array of publications, many of which are available through their web sites or through colleges' and universities' collections.   The research findings are both contemporary and easily understood; they're written principally for practitioners, who generally prefer summarized, straightforward information. Therefore, both beginning and more advanced students are able to access and understand these resources.

Of course, as sources for reading go, there is always the "wonderful world of the Web." Currently, many government agencies have their own web sites, and a review of them can give insights into subject areas being researched.

However, writers must be careful when accessing information from Internet sources; some are credible while others are not. Please refer to "Step Three," "Evaluating Sources," where we provide suggestions for determining the credibility of a source. Additionally, your instructor and librarians can provide assistance in source evaluation.

READING RESEARCH

Beginning writers often are not selective about choosing which resources to take out of the library. They tend to copy faithfully everything they have read without considering its relationship to the topic and other gathered information. And somehow, they equate the paper volume of information with having gathered enough information to begin to write.

While it's gratifying to find the research material we were looking for, the writer has to understand what each source means. The content certainly communicates information, but it also gives the writer a clear picture of whether the topic is still too broad, would require the writer to have previous understanding of the topic, or would require the writer to give extensive background to the readers.

Furthermore, without understanding the content, writers have no idea whether it relates to our ever-narrowing topic, nor do they understand its relationship to other sources. Do they agree, contradict each other, or duplicate the same idea?

READING TIPS

- Survey the source before reading. Look at the table of contents, any information listed about the author, and, if applicable, the summary or abstract directly under the title that will tell you what the article is about.
- Note the author's credentials and look at the publisher of the material.
- Read the material quickly the first time to get a general idea of content and format. Do not highlight or underline.
- Read the material again, this time slowly; underline or highlight <u>sparingly</u>. Often writers mark too much so nothing really stands out.
- Annotate (make notes about content) in the margins so you can later find the information in the material. (Only do this if you have made a copy of the article or own the book please!!)
- Write a summary of the article content, 2-3 sentences either at the top of your copy or in a small research notebook, and give the thesis statement
- Collect all bibliographic material you will need to reference the source in your paper.
- Identify the source as a primary source (the document itself) or a secondary source (something written about the document itself).

**This information will be valuable later when we discuss evaluating resources in "Step Three."

INTERVIEWS

A review of the written materials and scholarly research on a topic is usually the best place to get your information before you begin research. Most instructors give paper assignments that are based solely on academic research, however, sometimes an interview can be an important addition to the information-seeking process. An interview is a structured conversation. It can be most useful for obtaining information and/or direction to the most current information on a topic after initial reading has been done.

Before you plan to conduct an interview, you must first make sure you have the permission of your professor. At most colleges and universities, conducting any type of research requires obtaining permission from the *Institutional Review Board* (IRB), which is made up of a panel of people, usually faculty, who review all research proposals to make sure that the rights and interests of the subjects are protected, that the research is ethical, and that no harm comes to the subjects.

However, if you have permission, an interview also may be valuable for developing examples to support ideas gained from reading. Too, interviewing an expert in a specific subject may allow the writer to gain clarification on concepts that may not be readily understood when reading research, e.g., "tracking" versus "individual goals" in education.

Obviously, the value of an interview largely depends on who is being interviewed and the experience of the person doing the interviewing. Not everyone connected with a subject will be the right person to provide the best information on the topic, so writers first need to be careful that selected interviewees are knowledgeable in the field.

When selecting an interviewee from the community, the interviewer should be aware of possible subjectivity. For example, if a writer were preparing a research paper on "domestic violence," opinions and knowledge bases could be biased, depending on the interviewee's position within the community. For example, a manager who works in a women's shelter and who has come from a domestic violent past may be more concerned with obtaining immediate help for women, while an academic who researches domestic violence might be more concerned with finding similarities and differences among women who have successfully gotten out of violent relationships and stayed out.

Therefore, when interviewing, it is important to be sensitive to position bias. Also, those you may interview are not always familiar with the state of scientific research on a subject. Of course, practitioners do have a very good experience-based sense of the realities affecting a situation or how scientific findings actually apply in the "real world."

CONDUCTING INTERVIEWS

As an interviewer, a writer should conform to some time-tested principles of interviewing. First, the writer should not presume that the person identified for a possible interview is indeed the best source for information. Therefore, when the potential interviewee is contacted, the interviewer should be precise and clear about the topic of the interview and the information to be elicited. If the person confirms knowledge of the subject matter, the interviewer should set up an interview date and time,

20

ensuring at least 30 minutes for the interview. If the person claims not to be knowledgeable on the subject of the interview or for any other reason cannot be interviewed, an effective interview technique is to ask if the person can recommend someone else to speak with.

Once an appointment has been arranged, the next step is for the interviewer to prepare questions for the interview. A well-prepared interviewer can gain the respect of the interviewee and inspire quality responsiveness. Questions should be open-ended and fair. Open-ended questions are those that cannot be answered with a simple "yes" or "no." A question such as "Do you believe women who are abused could get out of the relationship if they wanted to?" should be recast as, "Why do you think women stay in abusive relationships?" The second question elicits a thoughtful, detailed response as compared to the first one that calls for only a one-word answer. As regards "fairness," asking awkward or inflammatory questions often begets a similar response.

An example of this type of questioning occurred recently when a well-known television personality interviewed a congressman suspected of involvement in the disappearance of a Washington, D.C., intern. Shortly after the nationally televised interview got underway, the interviewer asked bluntly of the congressman, "Did you kill Chandra Levy?" Whatever the intended result of the question, one thing it did not do was foster responsiveness.

Another potential interviewing pitfall is using leading questions, e.g., "Why do so many college students drink excessively?" or "Why are members of minorities disproportionately involved in drug use?" Note that each question assumes certain things, i.e., student alcohol abuse and drug involvement, which the interviewee may not agree with. Keep in mind that the interviewer's purpose is to find out what the interviewee knows, not misuse the occasion to express the interviewer's personal opinion or to influence the direction of the response.

INTERVIEW TIPS

While there are no lists of questions that all interviewers should use, here are a few general suggestions:

1. Review the questions you intend to ask before the interview to ensure they are clear or not too long or complicated.
2. Make sure that multiple questions are not embedded in a single one.
3. Make certain that questions are designed to gather biographical and background information about the interviewee to establish credibility and validity of remarks.

Before the interview begins, the interviewer needs to ask permission to videotape or audiotape the interview. If taping is approved, it is still a good idea to take notes on key points in case of mechanical failure and also later to help locate segments of the taping. Because an interview is a structured conversation, the interviewer should feel comfortable asking questions to help clarify information that seems confusing or incomplete. Otherwise, the interviewer who later becomes the writer will find it difficult to report on information that was not completely understood.

Asking follow-up questions is another important interviewing technique for gathering information. Often an interviewee may introduce an idea relating to the research subject that the interviewer may not have anticipated but now understands as an important trend of thought or to clarify what may seem to be contradictory information.

Practiced interviewers, after completing their listed questions, will ask, "Is there anything else that you think I should know?" This is done in the event the interviewee had assumed the interviewer already had knowledge of key elements of a subject because questions were not asked about them. Additionally, it gives the interviewee the opportunity to share information that may not be the focus of the interview but that may be closely related and helpful in providing perspective on the research subject or tentative thesis.

Obviously, the good interviewer will remember manners and thank the interviewee for the opportunity to speak, also requesting the opportunity to contact again should any clarification be necessary. Later, sending a copy of the final paper is a recommended practice that can be gratifying for the interviewee and also maintain positive relations for any subsequent interviewing or surveying that may be required (Williams & Brydon-Miller, 1997, pp.50-52).

SURVEYS

Surveying is another method for learning more about a topic, supporting a thesis, or discovering direction. Once again, conducting your own survey requires both experience in research methods and in many cases, permission from the IRB at your school. Your instructor will let you know if you need IRB approval. If you are not going to conduct your own survey, you can refer to the research literature to find survey's that have previously been conducted, to find out their results, which can then be applied to your research.

Surveying generally connotes a wide and objective search for information or opinions. A survey is often administered in a less personal fashion than an interview although surveys could be given during an interview. Very often surveys are conducted through the mail or telephone. They also can be administered via e-mail or by other electronic means, e.g., via the Web. Surveys are popular with many researchers because a large number of individuals can be reached at a minimal expense, depending on how the survey is administered.

In general, survey questions have to be clear, to the point, should not be biased or leading. The survey questionnaires should not take longer than 15 minutes to complete. Making the survey form attractive to the eye will have a positive impact on the response rate.

It is important to recognize that information gathered via surveying is only as useful as the manner in which it was collected. It is essential that bias be controlled to the maximum extent possible. This is accomplished through use of sound research design principles, covered in depth in basic research methods courses. However, while this writing supplement for sociology students is not intended to be a primer for research methods, there is one critically important aspect of methods courses that has bearing here: random sampling.

Randomness in surveying means that everyone in a given population has an equal chance of being selected for questioning. Use of a survey usually implies that at least 30 people will be questioned, a number selected for statistical purposes.

Thus, if we want to gain insights into critical issues of concern to domestic violence, the surveying is limited to the geographic boundaries of the region sampled. However, the value in this is that so long as all persons in a given area have an equal chance of being included in the survey, a random sampling method is used to select the survey participants, guaranteeing the findings can apply to that entire population.

As a case in point, two of the authors of this manual recently conducted a mail survey (with fax as a response option) of police executives from California and Pennsylvania. The questionnaire elicited "top research priorities." Questionnaire recipients were selected, randomly, from comprehensive rosters of both states' chief executives. Every third chief listed on the rosters was sent a survey form. By using this random method, we would be able to stand on solid ground if anyone were to question the survey results in terms of their representing both states' chiefs' overall perspectives. Interestingly, among the top four categories of research interest was an area categorized as "written communications/report writing." The underlying research question was "What can be done to improve the quality of police officers' written communications?" (Perhaps supplying them with a copy of this manual?) For the record, the other three categories of research interest were community oriented policing, technology, and ethics.

Surveying should be viewed as an extensive undertaking that often does not involve one-on-one exchanges with the respondent. Its value lies in the large numbers of persons it can poll. The decision to interview versus survey, or to do both, or neither, is dependent on the information gap and the available time/resources for completion.

The sample assignment used in this text for illustrative purposes is the type that could involve interviewing or surveying although the timeline for report completion might be too short for a survey or survey and interviews. Because of the brevity if this text, we have not provided exemplar survey questionnaires.

THE TENTATIVE THESIS STATEMENT

In "Step Three" we'll discuss the **tentative thesis statement**, our third structural element, which is the result of thinking and gathering information. This tentative thesis statement then directs future thinking about which informational sources to use, how they fit together, and which organizational plan will be used to develop the paper content.

Take a look at this flowchart on the next page to understand how everything discussed up to this point works together to help us build the paper's structure and guide later decisions.

Writing Assignment

Generate Ideas

Task   **Scope**   **Readers**

Identify Appropriate Ideas

Interview   **Research**   **Survey**

Analyze

Tentative Thesis

(Cooley, 1993, p. 13)

At this point we label it **tentative thesis** because writers will continue to modify the thesis as they continue to gather and assess additional information.

References

Cooley, M.E. (1993). *The inventive writer: Using critical Thinking to persuade.* Lexington, MA: D.C. Heath.

Williams, B.T., & Brydon-Miller, M. (1997). *Concept to Completion: Writing well in the social sciences.* New York: Harcourt Brace College Publishers.

STEP THREE:
PREPARING TO WRITE

In the previous chapter, we looked at **narrowing the writing topic** and **gathering information through library research and interviews/surveys**. At that point in the paper writing process, we were concerned with finding information relating to the topic rather than selecting information that supported a position the paper will develop.

Even if we began to research with an idea of what we hoped to support, such as the negative effects of daycare on six year olds, many of our sources may have focused on information about "daycare" or "six year olds." Therefore, the writer's challenge is to look at how research sources meet, that is, how they fit together to develop and support a specific position. The sentence in the paper that tells our readers the idea we plan to develop and support is called the **thesis statement**. Let's take time to define exactly what we mean by this term.

DETERMINING A TENTATIVE THESIS

The final piece of structure, **determining a tentative thesis**, can happen only after a writer narrows the topic and conducts some research to gather specific information. The **thesis** then is the writer's <u>reasoned</u> conclusion about how information fits together or the position that will be supported by research. This doesn't mean that there is one conclusion or correct meaning only. Instead, the writer's goal is to show to the reader that the writer's thesis is reasonable or valid, based on information, not opinion or emotion. This reasoned conclusion is contained in a sentence called the **thesis statement.**

Some beginning writers misunderstand the sequential process of **thinking about** and **narrowing the topic, finding research resources,** and then **determining a thesis statement.** Instead, these inexperienced writers first develop a thesis idea, reflecting their own opinions, and then go to the library to try to find information to support the thesis idea.

This misunderstanding of the process creates major problems. First, it leads to generalizations and somewhat unsophisticated thinking, reflecting bias instead of knowledge. For example, a pre-determined thesis statement might be, "Daycare is <u>bad</u> for <u>all children</u> under <u>six years old</u>."

Not only has the writer carved out huge areas to write about- all daycare and all six year olds, but, even more importantly, has substituted an illogical generalization for a specific thesis statement. Why was this done? Because the writer thought it would be faster to make the research support the thesis rather than look at research to arrive at a thesis statement. This approach of forming a thesis statement first and then researching most often results in faulty generalizations that can't be supported.

The second problem with determining the thesis and then looking for information that supports it is just that: the writer ignores research that doesn't support the thesis, the sort of "I'm not looking" approach that often results in invalid conclusions. Using our example, research exists to show preschool children in daycare may be more independent, may learn how to relate collaboratively, may become more self-sufficient than children at home, information the writer would have to ignore. **Therefore, our thesis statement needs to reflect accurately what reliable research resources will support**.

It's really not difficult to arrive at a **thesis statement,** but there are mistakes to avoid. Consider the list below of common errors writers make when writing the thesis statement.

1. The thesis statement announces the topic rather than states a position. "This paper is about new security measures in airports."
2. The thesis is too vague or broad to convey meaning to readers or to give specific support. "The average college student today is much more advanced than students twenty years ago."
3. The thesis is too specific or narrow to write about. "Seventy-two percent of freshmen on this campus brought their own computers to college."

Now that we know what the thesis statement is <u>not</u>, we need to consider how to "get" one.

SYNTHESIZING SOURCES TO ARRIVE AT A THESIS STATEMENT

If you recall, we started out our paper-writing process by analyzing task, scope, readers, then narrowed to a writing topic through some preliminary researching/interviewing/surveying. We may even have had some idea of what we thought we might find in our research, but that was based more on opinion rather than resources.

Yet, as we collect information about the writing topic, what we read changes our thinking. Our research process begins to shape us as we find a variety of informational sources that broaden our knowledge of the topic, identify current, often contradictory, issues within the topic that make us realize our initial **tentative thesis statement** is no longer valid or quite on the mark.

Using our thesis example of daycare and six year olds, we noted that the writer had used an opinion to develop the thesis statement, "Daycare is harmful to all six year olds." But through careful gathering of sources the writer may have discovered the following:

1. The term "daycare" can be applied to a variety of different childcare situations, differing in purpose, number of hours spent, identity of caregivers, etc.
2. The original term "harmful" could refer to a number of different elements including developmental, psychological, or physical factors.
3. There may be trade-offs between the benefits and negative effects of daycare for a particular age that may make a blanket statement about harm less valid.

Therefore, by gathering information and comparing/contrasting the information in our resources, we have the chance to rethink/reevaluate our tentative thesis statement. We call this putting together of resources **"synthesis" or "synthesizing" our sources.**

THE SYNTHESIS MAP

It's hard to find a method that allows a writer to look at all the sources together to take an inventory of the information that has been gathered. This process, called **synthesizing,**

31

allows the reader to see how sources "fit" together i.e., duplicate, contradict, build upon others, to determine which sources to hold on to and which no longer fit or are necessary.

There is, of course, the possibility of spreading out the source material on a large surface such as your living room floor or the dorm hallway…. But even spreading all material out would still leave the task of crawling around from source to source, looking at each source individually and collectively to understand the content.

An alternate method for **synthesizing** sources easily is to create a chart that allows us to look at what each source develops individually and compare/contrast with other sources. (**Note:** If the writer has followed the reading research suggestions in "Step One," much of the work for filling in the chart has already been done.)

EXAMPLE OF SYNTHESIS MAP

Source Title *Intimate Partner Violence*	Source Title *Restraining orders*	Source Title *Discussing Domestic Violence with Family Practice Patients*	Source Title *Domestic Violence and Physical Health Consequences*
Thesis Statement of source There isn't one cause, nor one cure	**Thesis Statement of source** Restraining orders don't always work	**Thesis Statement of source** 5-20% of patients have experienced domestic violence, but few discuss it with their doctor	**Thesis Statement of source** Abused women have more health problems than non abused women
Major Points	**Major Points**	**Major Points**	**Major Points**
1. Some of the causes are poverty, alcohol and traditional social roles	**1.** Prevention of Domestic Violence Act was enacted to address the fact that domestic violence is a serious crime	**1.** Domestic violence is common among women seeking treatment	**1.** Abused women have a 50% to 70% increase in gynecological, central nervous system, and stress-related problems
2. Cures can be the empowerment of women, education, and income	**2.** It is the courts responsibility to protect victims	**2.** There is no typical "victim profile"	**2.** The long term problems associated with domestic violence last even after the violence has ended
3.	**3.** Victims tend to ask to have a restraining order dismissed, but then need it again	**3.** Violence is always associated with fear	**3.** Abused women also had more gynecological problems than those who were not abused

SYNTHESIS MAP FUNCTION

In a sense, this chart provides a snapshot of all gathered information. The writer can then trace connections from source to source, making sure that similar ideas in fact are contained in each source. In chart form the writer has a map of everything "owned" and can make decisions about which sources to keep and which sources to discard because they no longer reflect the developing idea. The writer can then identify information gaps in remaining sources and begin searching for additional research.

But here is a surprising result of using the **synthesis map**: writers often find connections that they hadn't thought about before or realized existed. This discovery may lead writers to pursue an entirely different direction for the paper, scrapping one tentative thesis idea for another.

This reassessment of information and its connection to other information, leading to new understandings/conclusions, mimics how we think. We always add additional information or reassess what we know to draw different conclusions. How often do we and others say that we made the best decisions based on what we knew at the time, meaning that with later information we might have drawn different conclusions.

Perhaps a sad analogy to this point is the post "9-11" reappraisal of information. Surveillance reports and FBI requests, INS spotty monitoring of those in the United States with student visas, and the identification of a pattern of flight school enrollment had little individual significance, but, later examined together, the pieces had a distressing connection.

In short, the **synthesis map** allows the writer to make writing/thinking decisions about the future direction of the paper by performing the following tasks:
1. **evaluate** validity of sources and eliminate some sources
2. **identify** and resolve gaps/conflicts in information
3. **readjust** the tentative thesis to become the prime time thesis
4. **discover** a different or modified thesis statement

EVALUATING VALIDITY OF SOURCES

Most writers know that at some point they will have to sort through the stacks of photocopied sheets of research to identify "treasures" and put the rest in the discard pile. One of the first issues to consider when synthesizing sources is to consider the **validity** of the source.

Many writers might assume that **validity** only means that the information is reliable. Certainly that's important and increasingly more complicated with Internet research capabilities and advertising/promotional materials closely mimicking educational/scientific formats. Therefore, writers must examine each source of gathered material rather than assuming that it is valid because it is in print.
How does the writer evaluate the validity? Let's consider questions writers should ask themselves when looking at sources to determine **validity.**

DETERMINING VALIDITY

1. What do I know about this source? Is this source from an academic journal such as *American Sociological Review*, which is peer-reviewed or is the information from a magazine such as *Family Circle* or *Newsweek*? Is it from a reputable Internet source?

Internet sources, such as those from the *Centers for Disease Control* (www.cdc.gov), are more reliable than "Joe has a disease.com."

2. How current is the information given in the source? If it is less current, the possibility is greater that information has been changed/amplified since the original publication date. Therefore, it may no longer be valid in light of further research/information.

3. Is the authority cited in the source an authority in the field being discussed? Are the authority's credentials/degrees reputable such as an MD or a Ph.D? Did the information come from a newspaper that might not have reported the information accurately?

4. Does the reputable source carry a predictable bias, that is, maintain a consistent, predictable point-of-view because of the writer/publication's position? Think here about sources that have either a liberal or conservative political view, or a health magazine that predictably will promote the value of good diet and exercise. While this predictable bias doesn't mean the information isn't factual, we can expect that it will not carry an opposing point of view.

5. Am I getting conflicting information from sources? If this happens, conventional wisdom suggests going with the majority unless the dissenting voice is considered to be the major authority in the field.

IDENTIFYING AND RESOLVING GAPS IN INFORMATION

Because we have been so wise to consider information gaps now rather than discovering them in the midst of writing the paper (the day before it's due), we have the opportunity to eliminate and/or gather additional information needed to develop and support either the thesis statement developed previously or the new thesis statement that popped out when we synthesized our sources.

If we could list one of the major mistakes writers make at this point, it would be the writer who tries to "make do" with sources, attempting to stretch them to appear to support/develop ideas, when the reader realizes little support exists. Such writers either evaluate readers as less perceptive or lack the initiative/energy to go retrieve more information. Now, of course, if you're trying to write a paper for the next day's class....

Information gaps or need for additional information don't indicate we were bad researchers. Instead, as noted before, we may have significantly changed our intended focus (thesis statement) as a result of information gathered or discovered that we could not find the resource material we thought we could find. The latter is especially true when we are researching for very recent information. There may be a significant publishing gap between events/discoveries and the publication of that information.

ADJUSTING THE THESIS STATEMENT

Remember that we said the tentative thesis was a starting point, reflecting what we intended to support, based on preliminary research? Therefore, the tentative thesis continues to be modified, based on what we discover from additional reading and then synthesizing our sources.

In fact, writers often "discover" and pursue a new thesis idea, often much more narrowed, suggested by the gathered research.

THESIS DEVELOPMENT FLOWCHART

While books encourage us to believe that writing is a linear process, it isn't. Instead, writing is a recursive process, meaning we keep going back to readjust what we have already done. Look at the flowchart below, focusing on the thesis statement that seems to capture this idea of moving forward and backward.

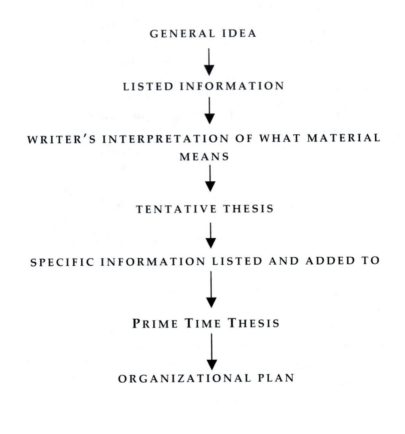

GENERAL IDEA

↓

LISTED INFORMATION

↓

WRITER'S INTERPRETATION OF WHAT MATERIAL MEANS

↓

TENTATIVE THESIS

↓

SPECIFIC INFORMATION LISTED AND ADDED TO

↓

PRIME TIME THESIS

↓

ORGANIZATIONAL PLAN

STEP FOUR:
DRAFTING

Many students (not our readers, of course) define **drafting** as the process of sitting at the computer for many hours, preferably after midnight, to begin and finish a paper due for class the next morning. Then, in order to make sure the paper is perfect, they might run a quick spell check before printing and dash to class, clutching their efforts that in daylight look considerably less brilliant than they did at 4:00 am. The bottom line is that haste does not produce quality. The express lane may be great for grocery store shopping but not for paper writing

As you may guess, the definition we will use for "**drafting**" will show significant differences from the process above. For us, **drafting** means the process when ideas are translated into words, sentences, paragraphs, and then revised to provide for clear communication to readers. And even more importantly, this process requires time to let writing "cool" before coming back to reconsider, rearrange, and revise.

 Therefore, the hyper-speed paper method many students try unsuccessfully to use won't work, in part because of our brains. Thinking and translating ideas into words is a higher-level brain function and looking at relationships among words in order to state information clearly is also a higher-level brain function.

Quite simply, our brains cannot perform both tasks at one time, needing to choose between the thinking stage and the writing stage at any one time. Yet, both thinking and writing are important to produce an effective writing product. We need to think logically and then to state our ideas clearly, guided by assessment of readers' "entry behavior": what do they <u>already know</u>, what do they <u>need to know</u> in order to understand, and what is the <u>best way to state information</u> for them. (Sound familiar? See "Step One.")

Our first step in the writing process is to focus on developing the paper's content by determining an organizational plan for writing.

DETERMINING AN ORGANIZATIONAL PLAN

When we choose a plan for organizing the entire paper, we make a selection based on how information will be best understood by readers. This differs from a content outline, reflecting the specific content of the paper. Instead the organizational plan is more an overview of how the paper will fit together. When we understand how the paper will be organized, we are then able to decide where specific information can be placed.

ORGANIZATIONAL PLAN CHOICES

There are general, overall organizational plans writers can use, but the choice to be considered in terms of the paper's **task, readers, scope**. In the academic world, the writer often selects an organizational plan unless a specific format to be followed is given. You must also remember that most instructors give paper assignments with clear guidelines regarding how they should be organized. Keeping that in mind, let's look at the general descriptions of organizational plans.

GENERAL TO SPECIFIC

This organizational plan is based on deductive reasoning where the writer begins with a general statement followed by more specific information (examples). Note that this pattern makes the main point immediately and then goes on to develop supporting details and assumes the reader will agree with the general statement.

A possible example here could be for a writer to state that efficient police call handling of domestic violence cases (general) can be traced to police intervention training (specific).

SPECIFIC TO GENERAL

This organizational plan is based on inductive reasoning where the writer starts with the specific and moves to the more general. Often this pattern is used to give the reader time to consider individual, specific pieces of information before the writer indicates what they add up to. This is especially useful when the reader may be unfamiliar with the paper's focus or may not initially be likely to agree with the general statement.

We unfortunately have recent examples of this pattern. Consider the attempts to determine who was responsible for mailing anthrax, killing several people (specific) with no one convicted of this crime (general).

CHRONOLOGICAL

This organizational plan uses time to determine the sequence of ideas. This is a common pattern for narration in fiction and history but often is <u>not</u> the most effective method for presenting more complex information requiring analysis.

This pattern could be used to list the legislative initiatives on domestic violence, focusing on the sociological issues in chronological order.

SPATIAL

This organizational plan uses physical location to determine the order information is given.

This organizational pattern could be used to develop a domestic violence hotline or to describe a crime scene.

CAUSE/EFFECT

Just as it sounds, this pattern focuses on what has caused something to develop a pattern, trend, etc. and then discusses the effects produced as a result.

For example, a writer might want to develop the causes of an increase in domestic violence in a certain location and then discuss the effects in certain identified areas, such as economic development, injury and death rate, deterioration of existing structures, increase in media coverage of domestic violence case, etc.

A SAMPLE ORGANIZATIONAL PLAN

Let's look at a sample tentative organizational plan to
answer the sample questions. To refresh your memory,
let's look at the sample assignment again.

Sample Assignment

The field of sociology provides a wide variety of research
opportunities, interesting to both students and
practitioners. Therefore, the purpose of this paper is first to
ask you to identify a research topic that provides recent
information to sociological practitioners that would be
helpful to them in their mission to find out about some
aspect of social life. To help you select a current topic,
please review current sociological publications and/or
interview some practitioners to gather possible ideas.

After you have selected a writing topic, prepare a research
paper that summarizes at least three key aspects of the
current sociological area relating to the aspect of social life.
The paper should be approximately ten pages, using at least
five different research resources. All resources should be
documented using ASA documentation format. The due
date is October 21, and the paper is worth 30 percent of the
course grade.

As an added incentive to perform well on this exercise, a
review panel of sociology professors will read each paper
to determine which three papers best address topics of
urgent value to social life.

A tentative organizational plan to answer the sample
assignment's questions follows.

- Thesis statement that reflects the effectiveness of increasing community awareness with the reduction of domestic violence
- Definition of domestic violence (DV)
- Summary of research on the criminal justice system and domestic violence
- Theory/practices that help prevent and assist with domestic violence cases
- Working with physicians
- Working with police

Show how lack of complete understanding of DV principles/practices listed above greatly reduces effectiveness in the handling of domestic violence cases.

DEVELOPING A WRITING PLAN

Once the writer determines the basic organizational plan for the writing project, the next step is to create a writing plan to determine how information will be structured in the paper. As writers follow the following steps, individual preferences emerge. Some writers, particularly those in the more technical fields, may find an outline essential to begin writing. Outline forms may range from a blueprint approach with each detail listed to a more general topic outline.

Other writers, who may not find an outline approach useful, begin with a general idea of what needs to be included but create a series of drafts in order to develop content and discover where material should be placed, using a cut-and-paste method to extract good material from all the drafts to arrive at a final draft.

Writers need to determine what works for them. In most cases there is not a significant difference in completion time, but the emphasis here is in finding a method that produces successful results for you. Interestingly, many writers still follow the rather arbitrary rules set forth by their seventh grade English teacher, never understanding that, within the writing process, many possible approaches exist.

Whatever process a writer chooses, the list below offers **general planning suggestions.**

1. Look again at reader analysis bullets ("Step One"), remembering that your goal is to communicate effectively with the readers.
2. Make sure you have completed the gathering of information and necessary thinking before you begin to write.
3. Have all the reference material at hand so you can provide required documentation for information to avoid plagiarizing.
4. Identify the sections of the paper you will need to support the thesis statement and develop ideas completely
5. Budget your time to write one section at a time, taking a break before moving on to write the next section.
6. Build paper sections that logically flow from previous sections to the next section.

PLAGIARISM

As writers begin to draft, they incorporate researched materials into the body of the papers, which raises the issue of plagiarism.

It would be hard to determine whether plagiarism ranks higher than others on the academic dishonesty scale, but it certainly attracts our attention. At the professional writing level, we are aware of the plagiarism controversy involving respected historians Stephen Ambrose and Doris Kearns Godwin. Both, noted historians, apparently used minor sources, undocumented, in their publications. In each case, a source was very familiar to the writer, and plagiarism occurred when Ambrose and Kearns used exact wording of the original source, incorrectly thinking it was the writer's own thinking.

In another case, Nebraska school officials lost a $250,000 grant to study the effects of humans on the environment and the relationship of insects to people. This grant was awarded to the school district at the request of an award winning science teacher. It turns out this teacher used major portions of a grant by another school district in her grant. This type of behavior not only ruins reputations, but hurts many of the people involved.

At the instructional level, we see web sites offering advice and software to detect plagiarism ("Just type in a sequence of five words…") competing with sites selling papers, offering advice on personalizing them to avoid detection such as *Plagiarism.org* and *Turnitin.com.*

So, what's the big deal about plagiarism? Quite simply, it's the attempt, deliberate or unintentional, to pass off someone else's work as your own. While some students do deliberately set out to be dishonest, most others do so by failing to acknowledge (**cite**) their sources or document sources correctly and, by default, plagiarize.

Paraphrasing is when you condense the meaning of what an author wrote and put the passage into your own words. This is OK to do, but it takes practice. Remember, if it's not your original thinking, you need to give credit (**citing**) to the original source. <u>No matter how you use the information--quotation, paraphrase, summary-</u> material <u>must be cited</u>.

Documentation does more for the reader than establish ownership. On a more practical level, documentation of sources provides a real service to readers. By providing references, the writer allows the reader to find easily the original material and read it in its entirety. In a sense, it's not too different from a web site link on our computer screen, alerting us to additional information.

STYLE MANUALS

Even though writers know that they must document sources, how then can they avoid unintentionally plagiarizing sources used in the paper? The best answer to that is buy a reference manual that discusses and gives examples of correct **documentation formats** for different types of sources, including books, periodicals, journals, on-line sources, interviews, Internet sites…

Before buying, the writer should determine which **style manual** is used in the discipline because formats will vary. For example, <u>Publication Manual of the American Psychological Association</u>(APA) or the <u>American Sociological Association</u> (ASA) are frequently used in social science writing. Instructors often list the required documentation format in the course syllabus or can offer correct purchase information.

WRITING DECISIONS: QUOTATION, SUMMARY, PARAPHRASE

Now that we know how to find the correct way to document all different types of sources, we need to decide how best to present the information in the body of the paper. While quotations, summaries, and paraphrases carry our researched information and require documentation, they are used somewhat differently.

QUOTATIONS

Quotations make writers comfortable. After all, the writer needs only to write the exact words, place them inside quotation marks, and give the source. In fact, some writers feel so comfortable using quotations that their papers seem to be made up of a string of quotations with very little writing to connect them. It's almost as if the writer has cut and pasted original sources in the paper.

While there's nothing like a good quotation that gets to the heart of the matter and states things so well, nothing is more deadly than reading a series of quotations that seem to be there because the writer didn't know how to extract the necessary information from the rest of the quotation. Generally speaking, if the source provides narration, as our example below, a quotation is not a good choice for sharing information with readers.

Look at the following magazine paragraph about the recent arrest of Jose Padilla, accused of planning to detonate a "dirty bomb" in Washington, D.C.

Padilla first surfaced in February when he went to the American Consulate in Karachi to get his new passport. The consulate, suspicious of a Hispanic with an assumed Muslim name in Pakistan, later flagged U. S. intelligence operatives, including FBI investigators. Not long afterward, American officials say, Pakistani authorities detained Padilla and a non-American associate for investigation on possible immigration violations, then let them go. (Hosenball, Hirsch, & Moreau, 2002, p. 32)

Notice the quote forces the reader to read more than is necessary. Writers who over quote rarely interpret the information; they just give it. This leaves the reader to try to extract the meaning the writer intended the reader should understand, woven into all those words. Expecting the reader to extract the necessary information is not a good writing strategy. First of all, there is a definite possibility readers will discover their own meaning that may not be the writer's interpretation. Secondly, reading quotations, especially those that shouldn't be quoted, can be deadly boring, encouraging readers to skim over content or stop reading altogether.

Are we trying to scare writers away from using quotations? No, but the point is that quotations are a bit like hot sauce: they get the reader's attention, often create a specific sensation, but, used too often or too much, the reader becomes irritated or immune to the effect.

Use a quotation when:
- the source is primary data(the actual document or piece of information that others write about)
- the quote is from a highly respected person or a recognized authority in the field

- the wording is so well stated or precise it could not be effectively restated
- the writer disagrees with the source and wants to establish fairly and precisely what was stated (Booth, Colomb, & Williams, 1999, p. 174)

Remember that, because we write for others, we must interpret or explain the quotation, always relating it to the points we make.

SUMMARIES

Summaries operate in somewhat an opposite direction because they <u>sum</u> up the important points of the original article in an abbreviated form. Most often summaries are used to provide background information or explanation to readers when the writer determines it is not necessary for readers to read the original source but to have access to <u>all</u> the information.

There is no exact word count for summaries although some texts suggest approximately one-fourth of the length of the original. Regardless of length, if the summary is to serve as a substitute for the original, writers need to be concerned that it:

- states the source's thesis statement
- is a condensed version of the original
- is objective, eliminating writer's opinion
- is complete, reflecting all major points rather than certain aspects only

PARAPHRASES

A paraphrase is another strategy to avoid excessive quoting and exists between the summary and the quotation. Like the summary, the paraphrase is stated in the writer's words, and, like the quotation, the paraphrase states each idea contained in the original source in approximately the same number of words. Paraphrases can be powerful because they allow the reader to give the detail of the original ideas without being locked into the language of the quotation. Often paraphrases can state information in words that readers may find easier to understand.

Yet there can be problems. How do you rewrite the material in your own words without either:

- creating awkward, strange writing
- **shadow plagiarizing**, that is, creating sentences whose word choice and sentence structure are so close to the original as to almost be the original source's "evil twin"?

The following suggestions should offer direction to writing effective paraphrases.

The writer should:
1. Rewrite material from the original source in approximately the same number of words.
2. Give an in-text citation for the source, using the discipline's style manual.
3. Keep those words from the original source that are so well stated a better choice doesn't exist, but put quotation marks around them.
4. Reflect the original tone of the source with verb choice, i.e., "complains," "praises," "explains" to place the paraphrase within the context of author's attitude toward the subject.
5. Look at the original source only when finished to determine if the paraphrase reflects the meaning

and avoided wording of the original (shadow plagiarizing).

6. Use quotation marks around any words from the original source. (Lester, 1999, p.112)

References

Booth, W.C., Colomb, G.G., & Williams, J.M. (1999). *The craft of research.* Chicago: The University of Chicago Press.

Hosenball, M., Hirsch, M., & Moreau, R. (2002, June 24). *Odyssey into the shadows.* Newsweek, 139(25), 28-35.

Lester, J. (1999). *Writing research papers: a complete guide (9th ed.).* New York: Longman.

STEP FIVE:
REVISING AT THE CONTENT LEVEL

The previous chapters offered guidelines for writers to generate ideas, evaluate their writing possibilities, find and evaluate research sources, and develop an organizational plan for the paper that should result in a final draft that will need minor revisions rather than full-scale reconstruction.

Because our brains can't revise content and writing at the same time, a writer performs two different levels of revision. This chapter deals with content revision, always performed before the writer examines writing at the sentence, word and punctuation levels.

THE NEED FOR CONTENT REVISION

You may be wondering why, after following all the steps, <u>any</u> revision is necessary. Isn't it logical to expect you now have a paper that is ready to be turned in? In an ideal world, the answer would be "yes," but, in the real world, even the most carefully planned paper will have weaknesses that the writer needs to resolve before finalizing the paper. In other words, a discrepancy always exists between the content of the planned paper and the actual draft, not because the writer has done something wrong, but instead, because of continued thinking as the writer drafts.

Good reasons exist for writers' need to make content adjustments to the draft, adjustments that cannot be previously "planned away." However, adjustment decisions in general do not require the emergency thinking demanded of writers looking at the results of their papers where no planning or serious thinking had been done beforehand.

To illustrate the expected difference between the planned paper and the actual written draft, let's go back to examine our sample organization plan, developed in "Step Four," to answer the sample question. Remember that the focus of the question was on three key sociological aspects relating to domestic violence.

SAMPLE ORGANIZATION PLAN

PLANNING SHEET

Here is a copy of a sample preliminary planning sheet for answering the sample question.

Thesis statement- effectiveness of increasing community awareness in the reduction of domestic violence
- Definition of domestic violence (DV)
- Summary of research on the criminal justice system and domestic violence
- Theory/practices that help prevent and assist with domestic violence cases
- Working with physicians
- Working with police

Note: Show how lack of complete understanding of DV principles/practices listed above greatly reduces effectiveness in the handling of domestic violence cases.

WRITING PROJECT EVALUATION

Task-to discuss at least 3 key aspects of Domestic Violence from a sociological perspective (causes, prevention, victims).

Scope-10 pages- should be enough to develop it.
Readers- Instructor is the primary reader but other sociologist will be later readers. Discussion of features of DV to show instructor my mastery of material.

OUR DRAFT PROBLEM

Based on this planning, can we predict smooth sailing and that we will have little or no revision? NO.

Let's look at the potential problems of our sample plan, something that a student might not recognize until examining the draft. The good news, though, is that, with careful thought and planning, problems should be identified and resolved.

While our sample outline shows logical thinking and good planning, one may not recognize that the paper has potential problems. Note that our sample plan develops two separate strands of thought:
- Prevention of DV
- Police officers' misunderstanding of DV

While the ideas are interrelated, each could be developed as a separate paper, one devoted to prevention; the other focused more on police departments' misunderstanding of the issues involved in DV.

In fact, if we look at the first paragraph of a sample draft, we can see the problem already emerging. Take a look.

The purpose of this report is to identify and summarize key aspects of the domestic violence. This theme was identified after library work and a literature review was completed on the subject of DV. The literature review revealed that knowing the causes and prevalence of

DV is integral to prevention. No matter what the issue of DV, the research showed that there are effective means for dealing with DV after it takes place. <u>One of the biggest impediments to universal acceptance of preventing DV is that it is widely misunderstood</u>. <u>Few police agencies seem to practice have much training in the area of how to handle the cases properly</u>

In sentence 1, the student writes she will "identify and summarize key aspects of DV" and then in a later sentence states that the issues involved in preventing DV are less effective because its principles are misunderstood. Because each of these strands could be a separate paper, the opening paragraph seems to promise more than can be developed in a <u>ten-page</u> paper.

PREDICTING THE SAMPLE DRAFT

Using the sample writing plan, we could predict writing sections of the draft would look something like the following.

Section 1-thesis statement

Section 2-discuss/define DV (This could take several paragraphs to several pages). At this point his writing plan indicates he will write about <u>DV in general</u>, and even if he devotes several pages to it, he will be writing about it at a general information level rather than developing specifics.

Section 3-discuss a summary of the research on the criminal justice system and domestic violence. Here's where much of the difficulty begins. The sample plan shows that the writer will discuss what officers don't do to implement

effectively the <u>very aspects the writer has yet to discuss.</u>
Therefore, the paper's logic is beginning to break down.

Section 4-discuss theory and practices that help prevent
and assist with domestic violence cases. If your hair is
beginning to stand on end when you look at this section, it
should. Look at everything that would have to be covered:
causes of DV such as poverty, alcohol, social roles, and the
cures such as empowerment of women, education and
income.

To summarize, because the writer of this sample draft
probably will not recognize the preliminary problems
created when the writing organization plan was developed,
the draft will try to develop the link between DV and
criminal justice system, police department
misunderstanding of DV cases, and how proper training
can increase the effectiveness of handling each case. This
leaves the writer with a draft that will be able to hit only the
surface of each area, supported by general information
only.

POSSIBLE SOLUTIONS TO THE DRAFT PROBLEMS

While the writer of this sample draft may feel panicked by
the results, she can still develop a good paper from this
shaky start if she diagnoses the problem as the too broad
focus, implied in the first paragraph, and narrow it. She
should consider:
1. the match between what is written and the
   assignment's writing task
2. which subject area best meets the assignment task
3. how to eliminate/narrow ideas within each area-
   causes, cures, and prevention
4. a modified **thesis statement** that shows what will
   be supported in the paper

The point we are making is that the draft reflects our thinking **at a particular point**. Later, when we see evidence of our thinking in our draft, we have the opportunity to reevaluate and revise as necessary. It's an expected part of the writing process, one that, no matter how competent we are as writers, can not be skipped.

THE PLANNED PAPER VS. THE ACTUAL DRAFT

In our sample paper, the draft problem involved rethinking the entire paper, and while serious, we saw that it could be solved. Even a draft that has a strong thesis statement can have problems in content development.

TYPICAL DRAFT CONTENT PROBLEMS

The following list identifies typical draft content problems that the writer must solve before finalizing the paper. Some of the more common problems are:

1. Some sections of the paper seem weaker than others.
2. The required length is longer than what you have written.
3. Gaps in information exist, forcing the reader to guess about missing content.
4. The paper develops an idea originally not intended or is over weighted toward a particular idea.
5. The thesis no longer seems to reflect the developed content.

POSSIBLE SOLUTIONS FOR DRAFT PROBLEMS

WEAKER SECTIONS

It is natural to expect that some sections of the paper will be stronger than others, either because you had more gathered information to develop the section or found it easier to write that particular section. However, the writer needs to be concerned when a section is obviously underdeveloped compared to others or lacks specifics needed to develop/support it. That's when the writer needs to evaluate the cause of the section weakness.

One possible cause for a weak section is that it makes a very minor point, or even an unintended one without a strong connection to the paper's main content ideas and could be eliminated from the paper without eliminating necessary content. If the information from this underdeveloped section is reflected in the thesis statement, the thesis statement will need to be revised so that readers don't expect to find the information somewhere in the body of the paper.

On the other hand, the writer has a more serious problem if the problem with a weaker section stems from lack of necessary information to develop the section, or the writer doesn't understand how the information fits together and has been deliberately vague, hoping instead that the reader already knows the answers. In this case, the writer has no choice but to provide more information by conducting more research and understanding it. To eliminate points because it's easier for the writer will produce a seriously flawed paper.

LENGTH PROBLEMS

Writers in this situation often make the worst possible decision: they decide to pad the paper, repeating general information over and over, adding extraneous information, hoping to add required pages without adding valuable content. This padding seldom lends itself to brilliant observations, producing about the same result in the reader as being put on "hold" on the telephone, forced to listen to pre programmed Muzak.

Therefore, the best first step to avoid attempts to try to make less look like more is for the writer to look at the assignment to make certain no part of it was left unanswered. If certain parts have not been addressed, the problem is fairly easy to resolve by answering the remaining questions.

On the other hand, if the writer evaluates all questions have been answered but the paper is still too short, other options exist. One possibility is for the writer to examine the paper, noting areas that seem underdeveloped or contain rather general content to make it more specific. Another tactic is to examine the examples and supporting information to determine whether adding additional support would strengthen (and lengthen) the paper.

If all points are developed and supported with adequate information, the writer needs to focus on the thesis statement, determining whether an additional, relevant idea should be added to it. In other words, the paper may be less than the required length because the thesis statement excludes some content areas needed to answer the question.

GAPS IN INFORMATION

The most obvious approach here is for writers to look at the gathered research to determine whether the material for whatever reason was not included. If, however, material needed to fill in the gaps doesn't exist, the writer will have to continue gathering information

PAPER TAKES ON DIFFERENT DIRECTION

Sometimes the draft turns out to be very different from what the writer thought/planned. In some cases it could be that a minor point in the original plan now becomes the draft's major focus. In other cases, the paper has drifted away from the intended thesis to develop something quite different.

In both cases, when the discrepancy exists, the writer needs to distinguish whether the draft is very off-focus or whether it is a better idea for the writer to develop in the paper. If the "new" content is merely getting off the subject, then the writer will probably need to eliminate it. However, if it seems like a better idea or a more focused idea than the original, the writer will want to revise the thesis statement to reflect this new, improved direction.

THESIS STATEMENT PROBLEMS

(See discussion above.) Frequently, a writer has to adjust the thesis statement after reading the draft. This new thesis statement, which we labeled "Primetime Thesis Statement" in the flowchart at the end of "Step Three," is a logical extension of the writer's evaluation of the draft. Very few writers can avoid making even minor changes to the thesis statement, necessary to reflect the revised, drafted content.

WHAT DOES REVISION SAY ABOUT ME?

Writers need to move beyond the point of ego, thinking that they, among all others, have mastered the art of perfect writing on the first try. Literary history is filled with stories of professional writers who struggled to produce what became masterpieces after many revisions or were forced to abandon the entire project because it couldn't be brought under control to meet readers' needs.

Content revision is a necessary part of writing. As we continue to change our perspective or reevaluate on the basis of additional information, we also need to readjust our drafts to reflect these content changes. Often the draft, a written record of our thoughts, looks less clear than what we thought we had planned for, requiring us to add and delete, reorganize information, tasks our brain can't do perfectly in one draft.

REVISION AT THE PARAGRAPH LEVEL

Writing for readers carries the responsibility of conveying information in a pattern that they can easily follow and

61

remain interested in what we are developing. While content development has been the major focus of this chapter, writers also want to look at paragraphs that should develop one idea, stated in the topic sentence.

FUSED PARAGRAPHS

As a reading cue, the topic sentence serves as a guide to the reader's eye about what information is contained within the paragraph, making it easy for the reader to find or refer back to information when re-reading the material. Therefore, all information in the paragraph should relate to the topic sentence; any other information should be eliminated or moved to another paragraph to avoid reader confusion.

The common writing term used to describe paragraphs that develop more than one idea is "fused paragraphs." If your instructor has ever written on your paper, 'Too many ideas in this paragraph," you weren't being complimented on your ability to think. Instead, you had fused paragraphs. Look at the example below where the underlined sentence shows how a new idea has been fused to the paragraph.

Technology has improved not only the way our hospitals function, but also our health. Doctors are now able to perform surgery without making incisions, thereby speeding up patient recovery time. Medical testing has also become more accurate. Doctors are able to diagnose problems often before the patient has symptoms, and computers now store patient records that can be accessed rapidly by many different medical personnel. Some major hospitals have telecommunication links with other hospitals to provide additional information about patients seen at other hospitals. <u>How people entertain themselves has also changed because of technology</u>….

Obviously, the underlined sentence shows a shift in direction and the problem resolved by shifting the sentence into its own new paragraph and then adding supporting details. But, since it was so easy spot the shift in thought, is it really necessary to "de-fuse" paragraphs? YES. Keep in mind that readers have not had the opportunity to follow you through the idea generating stage, the search for research, the organizational planning. The first time they know what you are thinking about is when they read your final draft.

 If paragraphs are the basic building block of the paper, then you have to be careful that they don't contain different strands of thought, seldom developed, having a vague connection to other material that only the writer understands. This, in turn, confuses the reader and dulls the original point of the paragraph. Why is that your problem? For three reasons:

- you are writing to communicate to the reader
- reading comprehension drops dramatically when a reader is confused and has to reread.
- bored readers often become non-readers (except for your instructor who is a captive reader, but then, most of your professional writing will not be done for an instructor.)

PARAGRAPHS WITHOUT SUPPORTING INFORMATION

Paragraphs that lack enough supporting detail should be the ones writers identify when they identify gaps in information in their writing. These problem paragraphs seem to hint at so much without giving adequate supporting detail, almost challenging the reader to fill in the missing blanks. Sometimes the information is omitted because the writer fails to anticipate readers' needs or skips over the details either because the writer doesn't understand the connections or lacks the necessary research to supply details.

CONTENT CHECKLIST FOR CAREFUL WRITERS

College juniors, enrolling at Pennsylvania State University at Harrisburg, took the Individual Writing Profile (IWP), a writing assessment tool designed to determine whether incoming students had the necessary writing skills to complete senior-college level assignments. All IWP writing samples were read and evaluated by two college instructors form outside the university and evaluated, using a Penn State-specific rubric, adapted from *Pennsylvania Writing Holistic Scoring Guide.* (Schmidt & Brode, 1994). The rubric sets up specific writing standards in seven different writing areas. The content standards listed below could serve as a quick checklist for writers to use when evaluating draft content.

CONTENT STANDARDS
FOCUS
- sustains a single point of view
- Answers the question

CONTENT
- Ideas are analyzed rather than named/listed
- Information and details are specific and relevant to the paper
- Paper develops a clear thesis and effective conclusion

ORGANIZATION
- Main points follow logical order or recognizable sequence
- Each paragraph deals with only one subject
- All paragraphs relate to the topic
- Logical transitions exist between sentences and between paragraphs

REFERENCES

Schmidt, J. H., & Brode, K.L. (1994). *Individual Writing Profile: Evaluative Criteria.* Unpublished manuscript, Pennsylvania State University at Harrisburg.

STEP SIX:
REVISING AT THE SENTENCE LEVEL

The previous five chapters have focused on developing the content of a paper so that it moves along logically with appropriate supporting information. In this chapter, however, we turn the focus to the sentence level to determine if our ideas are expressed clearly to avoid reader confusion about content and prevent misreading.

Sometimes after struggling to refine the thinking of their papers, writers are distressed that their drafted sentences look so—well, awkward or juvenile. They think, "The content is complex, so why do my sentences look so poorly written?" These writers have forgotten what we said earlier about the brain not being able to think and craft sentences at the same time. Consequently, we can expect the paper's thinking may be clear and precise, but the actual wording of those ideas in sentences will always need some revision.

REVISION ELEMENTS AT THE SENTENCE LEVEL

If you are mentally groaning, thinking this chapter is about to launch into the usual review of sentence mechanics and punctuation, you will be disappointed that is not our direction. Instead, we're going to focus on revision techniques that help streamline writing to make sentences more smooth, precise, and less wordy <u>after</u> we remind you about the importance of sentence structure and punctuation.

THE TWO MINUTE MECHANICS REVIEW

That we are not going to discuss mechanics and punctuation in no way indicates we think they are unimportant. But we know that so many other sources, probably including the textbook you used for Introduction to Composition, do such a thorough job in offering writing mechanics explanations and examples that we could not do the subject justice in so few pages. In case you no longer own that writing manual, many colleges have web sites, usually linked to learning centers, offering similar information for your reference. For example, Purdue University's web site (<u>www.purdue.edu</u>) has excellent resources for writers.

Because we would be negligent in not even mentioning standard correction areas to evaluate writing at the sentence level, the following chart lists them for you.

Correction Areas
Sentence Structure Complete sentences, avoiding fragments and run-ons Parallel construction Variety of sentence structures and sentence length
Writing Conventions Correct capitalization, punctuation and spelling Correct usage of pronouns, subject-verb agreement Avoidance of commonly mistaken words

ENSURING READABILITY THROUGH REVISION

While writers want to make certain that sentences are grammatically correct, they are aware that other revisions may be needed to ensure **readability.** This term means the reader's ability to read through a document without having to stop to reread confusing sentences or provide apparent missing links in the development of ideas.

You might be inclined to think it's really not that much of a problem. After all, the writer did all the work so can't the readers contribute something or use their imaginations? It's not quite that simple. If readers have to try to figure out what is being communicated, several outcomes can occur:

1. The readers' interpretations of the material may differ from the writer's intended meaning.
2. The readers' levels of reading comprehension decline after their eyes stop moving forward, having to go back to reread.
3. The readers may lose interest and stop reading altogether.

To ensure readability, we're going to look at four areas of revision: providing transitions, varying sentence pattern and length, choosing audience-appropriate language, and eliminating unnecessary words.

PROVIDING TRANSITIONS

Transitions, as you know, provide clues to the reader about the relationships between words and between sentences. By including well-chosen transitional words and phrases, the meaning is made clear, avoiding reader confusion or time to puzzle out the connections.

Categories of transitions exist, conveying different relationships. If your repertoire of transitions only includes, "in addition," "later," and "on the other hand," you will want to consult complete lists for those that convey very specific relationships among material including those that

- link similar ideas or add an idea
- limit or contradict an idea or show contrast
- indicate cause, reason, or result
- indicate time or spatial relationship
- indicate a following example
- indicate a comparison

Notice that the underlined transitions in the following sentences clarify sentence relationships.

John left his car running when he went into the store to pick up his dry cleaning. <u>Consequently,</u> the teenager who had just shoplifted several items from the convenience store jumped into John's car and sped out of the parking lot.

Officer Martin Carr is the new police chief in Chatham. <u>Previously,</u> he had been the deputy chief in Clearstream.

VARYING SENTENCE VARIETY AND LENGTH

When we write, we choose from 4 basic <u>sentence patterns,</u> each pattern ideally best suited to carry certain kinds of information. Understanding this correlation helps writers create less awkward sentences and increase reading comprehension.

Listed below are sentence patterns and the kind of information appropriate to each one.

1. <u>Simple sentence</u>-carries little information

The first study showed inconclusive results.

2. <u>Compound sentence</u>-carries two independent clauses whose information is equal in importance

Police recruits are trained in defensive techniques, and they learn domestic violence techniques to relate to the community in a non-confrontational manner

3. <u>Complex sentence</u>-carries one idea clearly subordinate to the other

Because the car wouldn't start, John missed the 7:00 am train.

4. <u>Compound-Complex</u> sentence-carries equal and subordinate ideas

After Professor Smith finished seven years' of research for his book, it was published, and the College immediately promoted him to Department Chairman.

Writers also need to consider sentence length as a factor to ensure readability. If you are interested in pursuing this concept, you can find formulas that show you how to calculate readability rating of sentences, paragraphs, or entire papers. For our purposes, we will follow a general business guideline, indicating the average reader can comprehend meanings in sentences averaging 15 words per sentence. For us, as sentence writers, we will have some sentences with word counts shorter than 15 words and others of more than 15 words.

Why not make all sentences 15 words if that is the best length? Several answers to this question exist. First, while you may not consciously be aware of it, your brain registers repetition in language patterns so that the reader becomes bored with the repetition. We could compare this impression to hearing the same song played endlessly or eating the same food items at each meal. We become dulled by the experience.

Secondly, writers must consider the information the sentence will carry as a determining factor in optimal sentence length. If we are conveying more complex information, using less familiar terms or words of several syllables, the information must be carried in shorter sentences to allow the reader to process the message. When we need to convey dense content, we are wise to break the information into a larger number of sentences rather than one longer sentence because of reading comprehension.

If you doubt what we are saying, consider the last time you needed technical advice to troubleshoot a computer problem and called the help desk. After the attendant used "jewel case," "install drivers," and "IPO address" in the first sentence, you knew you couldn't put all the information together.

CHOOSING AUDIENCE-APPROPRIATE LANGUAGE

We have already discussed the importance of knowing who our readers will be, allowing us to determine which information will be used, where it will be placed, and, finally, how best to phrase material.

Often our research takes us to information written by authorities in the field, i.e. a medical researcher writing about advances in treatment for non-Hodgkin's lymphoma; an economist writing about Vroom's theory of motivation, a forensic scientist discussing new ways to evaluate trace elements of DNA. In those articles we often have to struggle to define for ourselves the more technical terminology, using dictionaries and cross-referencing from other sources to understand the content.

Our readers are often in a similar position of not understanding what we communicate to them. We can help by providing a means for understanding by offering a definition of terms and ideas, comparing the idea to something already familiar to our readers, or stating it in different terms. For example, our medical researcher above might refer to "adrenalechtomized mice" for a medical audience while non-medical readers would better understand the term as "mice that had adrenal glands removed." Note that both terms carry the same meaning, but each conveys it in audience- appropriate language.

While this seems an obvious example, be aware that all disciplines have their own terms and often create "new" words, often not in the dictionary, generally used only within the academic discipline or profession. We call such specific words or abbreviations **jargon**, and, as writers, we have to be aware that jargon does not carry the same meaning to readers outside the discipline/profession.

ELIMINATING UNNECESSARY WORDS

Ironically, many people mistakenly believe strong academic and professional writing should not be understood by the average reader. They think the overly formal style of complicated sentences and elevated, often artificial, word choice, a sort of "showing off what you know," is necessary to reflect the importance of the content. Of course, the reader then needs to be willing to decode the message into easily understood information.

Fortunately, the newer emphasis on academic and professional writing is to make it accessible to its readers clearly and directly with a cleaner style, each word necessary for meaning. When writers use more words than necessary to convey information clearly and completely, they risk burying the meaning inside an avalanche of words, completely nullifying the original purpose for writing.

Perhaps writers should imagine they are charged twenty-five cents for each word used in the paper. The question they should ask about each sentences is "Did I get my money's worth?" If you look at information developed in the following sections to eliminate unnecessary word use, you should be able to say "Yes!"

REDUNDANCY

Redundancy is word choice that repeats the same idea. Probably all of us have heard the phrase "revert back," redundant because revert means "to go back," or heard advertisers warn us to order now to qualify for a "free gift," again ignoring the meaning of something given to, not paid for.

For your edification or amusement, can you identify why each of the following is redundant?

close proximity
biography of Ted Williams' life
12 noon
12 midnight
shorter in length
summarize briefly

We can also have redundancies within sentences where we repeat the same idea.

Pennsylvania moved to allow voter registration to drivers when they renewed their licenses. This was designed to encourage <u>unregistered voters who had not voted</u> to become involved in the election process. (Unless voter fraud exists, can't we assume this to be the case?)

PASSIVE VOICE

Verbs have two voices: active and passive. The active voice in generally preferable to the passive voice with a few exceptions that will be discussed. In active voice, the subject performs the action while in passive voice the subject is acted upon. Passive voice will always increase the number of words used.

The teacher <u>drove</u> the car. (active voice) 5 words
The car <u>was driven</u> by the school teacher.
(passive voice) 8 words

Notice that we have increased the number of words, changed the sentence subject from "teacher" to "car," and diminished the power of the verb to "was driven." Additionally, consistent use of passive voice is a guaranteed cure for insomnia because of the longer, dull sentences it produces. Readers have to be very curious to read through papers where the writer frequently uses passive voice. We offer the following paragraph as evidence.

The motion to approve the requirement that front doors be painted red by the occupants was passed. (passive) This bill had been sponsored by Martha Gladstone, owner of Ye Olde Paint Shoppe, and County Historian.(active) It was proposed last year but was defeated by a 7-3 vote against the provision.(passive)

Not only is the paragraph less interesting, it also leaves out crucial information. Who performed the actions? Who approved the motion? Who defeated the proposal last year?

Contrast it with this example in active voice.

The Commission on the Professionalization of Corrections Officers (CPCO) decided to continue discussion of increasing...." Here we know precisely who did what because active voice provides that information.

In short, passive voice increases word count and generally dulls sentence meaning. However, there are specific instances when writers consciously choose it.

1. When the actor is unknown or unimportant. (A 911 call, reporting a possible homicide, was received at 12:42 am.)

2. When identifying the actor would assess blame or liability.
   (A mistake was made in incorrectly flagging your account as overdrawn.)

COMBINING SENTENCES

At the draft stage, sentences exist that don't carry enough information to warrant "spending" the words in weaker sentences. To eliminate this problem, sentences can be combined to make a stronger sentence or the content can be embedded within another sentence.

Embarrassingly, we have all written sentences like the following, short and choppy with a "Dragnet" feel to them.

John went to the registrar's office. He couldn't schedule classes. There was a problem with his financial aid. (18 words) Notice the sentences are short and connected only because the <u>reader</u> carries over the information to the next sentence.

But, by combining sentences, we slightly reduce word count and more clearly show the relationship among sentence ideas.

John went to the registrar's office, discovering he couldn't schedule classes because of financial aid problems. (16 words)

In other cases, such as the example below, the writer may first need to look at sentence groups to analyze the sentence relationships before writing. The following sentences still have that disconnected feeling and we get duplication of ideas, "confused" and "didn't make sense" as well as word repetition.

The anthropologist was confused when he looked at the mummy. The death mask on the mummified remains of a laborer didn't make sense. Jeweled death masks were for pharaoh use only. (31 words)

Combined, a possible revision could combine three sentences into one sentence that is fairly long.

When the anthropologist looked at the jeweled death mask, used only for pharaohs, on the mummified remains of a laborer, he was confused. (23 words)

It could also be combined into two sentences, clearly reflecting the relationship.

When the anthropologist looked at the jeweled death mask on the mummified remains of a laborer, he was confused. These masks were used only for pharaohs. (26 words)

END NOTE

In some ways, *Six Steps to Effective Writing in Sociology* seems equivalent to touring Europe in seven days: we have hit the major points of interest and, along the way, have pointed out where you can find more information on specific areas in writing.

Our goal in writing this manual was to develop a process for writing that is logical and efficient. Yet good writing requires practice and a willingness to keep experimenting before achieving successful results, so just reading it will not make you a better writer.

We hope, too, you realize that all writers continue to struggle with the results we produce, constantly rethinking, rewriting, and revising what is on our computer screen or tablet, in order to make our content meaningful to our readers.